The Birds of the Oyster Coast

Text
by
Mike Gould

Original watercolour illustrations
by
Mike Roser

Produced
by
Melvyn Crow

This book is dedicated to the first twenty-five years of the
Whitstable Natural History Society

Published in 1998
by
The Whitstable Natural History Society
in association with
The Whitstable Improvement Trust

Printed in Great Britain by Oyster Press, Whitstable.

All rights reserved. No part of this publication
may be reproduced, stored in a retrieval system,
or transmitted, in any form or by any means,
electronic, mechanical, photocopying, recording
or otherwise, without the prior permission of the publisher.

© Whitstable Natural History Society

ISBN 0 9508924 1 6

Acknowledgements
Geoff Burton. Kent Ornithological Society. West Kent Recorder.
For assistance in compiling the checklist.
Anthea Skiffington. Canterbury RSPB Group Secretary.
For help with proof reading.
John Puckett. For final proof reading.

Foreword

I am neither an ornithologist nor a "twitcher", but like many others, I diligently feed my garden birds, and take pleasure in rare sightings.

It was on one of my frequent walks along the Tankerton front towards Herne Bay that I spotted a bird I didn't recognize: sparrow-sized, it was grey above, the breast was a sort of beautiful orange buff, and it had a distinctive white rump in an otherwise black tail, very marked in flight. As it scurried along the sea wall, it flicked its tail like a wagtail.

I couldn't easily identify it from my bird guide, and people I asked were not certain. Eventually, I concluded it must be a Wheatear - nothing to do with wheat or ears, but named after the Old Saxon for "white rump".

Since then, I have seen many other species along that seashore - Skylarks, numerous types of gull, the comical Turnstone, Ringed Plovers but also others I cannot identify. I am told that occasionally a Kingfisher may be seen, as well as pipits, stonechats and buntings.

This is where I am sure this excellent guide will be of great use not only to amateur enthusiasts like myself, but also to the regular birdwatcher. A book to have with you on your walks, it provides an instant, finely illustrated pocket guide whenever you ask yourself: what was that bird? It is just the sort of guide I have been searching for.

Dr David Heald

Introduction

Throughout the year, a visit to the Whitstable coastline is a thoroughly enjoyable experience, with many points of interest both historical and natural. Although, at a casual glance the sea and the mudflats may appear devoid of life, through binoculars a whole host of bird activity can be observed.

This book has been written to introduce the casual observer to the many birds which he or she is likely to encounter along the Whitstable coast. The common species, for example, Greenfinch, Goldfinch and Starling have been omitted as most people are likely to be able to identify these birds. There are, of course, many other birds that visit our coast from time to time, with several rarities making one off appearances.

Included at the end of this book is a list of over two hundred and twenty birds which have been seen during the last decade within the area of the old Whitstable Urban District Council's boundary and also including the coastal marshes towards Faversham Creek.

Habitats
The geographical position, coupled with the rich feeding to be found in the Swale estuary make Whitstable's coast ideally situated to attract migrating shorebirds and wildfowl. During autumn and winter months, thousands of waders and ducks, mi-

Waders feeding along the Swale during winter *Photographer: Melvyn Crow*

grating south or overwintering, take full advantage of the varied food sources found along the coast.

At the eastern end, Long Rock, Swalecliffe, provides many different habitats: seashore, salt-marsh, grassland and freshwater from the Swalecliffe Brook makes this a very important area. Westward, the rough grassland areas of Tankerton Slopes attract Meadow Pipits and Skylarks, whilst the bushes and trees around the Castle Grounds are well worth watching, especially during early autumn when birds such as Pied Flycatchers occasionally pass through. From here towards Seasalter, much of the low tide flats consist of consolidated shingle, which does not provide easy feeding and wading birds can be rather scarce.

At Seasalter and further into the Swale Estuary, the soft mud provides a rich feeding environment for a huge number of birds. The importance of these muddy estuaries cannot be over stressed as they support enormous numbers of tiny invertebrate life, which constitute essential feeding for migrating and overwintering birds.

Roosting Sites
High tide signals many birds to seek a place to rest and most fly over to the relative peace and quiet at Harty salt-marshes on the Isle of Sheppey. There are exceptions, with the more remote beaches along our coast tempting Turnstones, Dunlin and Ringed Plover to roost in considerable numbers. The outer shingle arm at Long Rock is perhaps the most well known and important site, but it is continually threatened by increasing human pressure. At Seasalter, three or four hundred metres west of the Blue Anchor public house, the upper beach has encouraged a similar roost, with each site often sheltering up to two thousand birds. The beach at the western end of Seasalter Golf Club, just below Sherrin's Alley has also been known to hold smaller roosts of waders. It is essential that these high tide rest areas are recognized and left undisturbed, as there are few places along the coast where our birds can find a safe haven.

The Birdwatching Year
Throughout the year many different birds arrive on our coastline, some to breed, some to spend the winter and others on their spring and autumn migration.

Late July and August sees the start of the return migration, with Greenshank and Black-tailed Godwits exploiting the pools and mussel beds at Seasalter together with Dunlin, Redshank, Bar-tailed Godwits and Turnstones. Many still sport their colourful breeding plumage. At this time most are on passage and stay only to feed a few days or weeks before continuing south to their winter quarters. In late September, the first Brent Geese arrive, followed by flocks of overwintering Dunlin and Knot during October and November.

Occasionally, strong north westerly winds during September and October herald the beginning of exciting seabird movements, with the North Kent coast being ideally placed to witness these spectacles. The second weekend of October 1997 was one of these special times when Leach's Petrels, Great Skuas, Arctic Skuas, Pomarine

Long Rock, looking towards Herne Bay *Photographer: Mike Gould*

Skuas, Long-tailed Skuas, Sabine's Gulls and continual Gannet movements were recorded. Most of these birds are observed at a range of a mile or more and telescopes are required to clinch their identification, although many approach close enough to be recognized using binoculars.

During the cold winter months the Swale mudflats support thousands of waders and wildfowl feeding on the many different organisms found at various depths within the estuary mud. To exploit these food sources they have evolved different bill lengths and shapes. Curlews use their long bills to reach deep into the mud for ragworms, whilst the Shelduck's bill is adapted to sieve the mud for tiny molluscs. Brent Geese sustain themselves on vegetation particularly eel grass which grows extensively on the Seasalter mudflats. Towards the end of March most of the overwintering birds have left our coast and are on their way north to their breeding grounds.

The first spring migrants start to arrive at the beginning of April; Wheatears, Yellow Wagtails and Swallows are some of the many seen moving westwards along the coastline. Over the sea, small groups of Whimbrel, Sandwich and Common Terns can also be observed tracking west into the Swale. May and June are very quiet months with relatively few birds to be found on the sea or mudflats. Only the odd pair of breeding Ringed Plover, Oystercatcher or Redshank may make an appearance. Mid-July sees the return of the first migrants, usually non-breeding birds, which once again starts the seasonal cycle of bird movement along our coastline.

Red-throated Diver *Gavia stellata*

Regular winter visitors, Red-throated Divers are usually seen flying past or loitering offshore, occasionally coming close to the beach. Sometimes up to half a dozen birds can be found in the Swale; however, generally they occur as single birds, sheltering during severe winter weather. Expert fisherman they regularly dive, chasing fish and swimming long distances underwater.

Large birds (53-59cm) they have dark grey upperparts and hindneck, the throat and face a contrasting white. Their red throat is only seen during the summer breeding period from April to September.

When swimming, the head and slender bill are often held at an upward angle, whereas the bills of the similar Great Northern and Black-throated Divers are much heavier and held more horizontal. Their fast hurried flight reveals white underparts, thin body, long neck, narrow wings and trailing legs.

Breeding occurs in the north-west of Scotland and beyond into the Arctic tundra. They make their nest on the edges of small pools and lakes, close to a ready source of fish and other invertebrates on which to feed. The legs of divers are positioned at the rear of their body which, on land causes them to shuffle along awkwardly. The nest is conveniently positioned at the water's edge and consists of a simple scrape, lined with vegetation and hidden amongst waterside plants.

Great Crested Grebe *Podiceps cristatus*

Great Crested Grebes are mainly winter visitors to the coast but single birds can often be found on the sea throughout the summer months. Most of the birds confine themselves to the Swale estuary, with a few other individuals venturing along the coast towards Herne Bay.

During cold periods, when inland lakes and reservoirs are frozen, the open sea offers them an alternative habitat for finding food. Here they have become well adapted to a marine existence and frequently dive for molluscs and crustaceans.

When swimming they can be identified by their dark body, long white neck and face, with a black crest on the rear of the head.

In past years, wintering flocks of five hundred or more birds have gathered off Shellness Point, and a flock of eight hundred and twenty one was recorded in February 1988. However, numbers have declined in recent years, with a sighting of only one hundred and sixty two birds in December 1997.

During strong westerly winds many of these birds find shelter off the Leysdown coast and are not seen in the Swale. In early spring they return to the lakes and gravel pits where they perform an elaborate courtship ritual of head shaking, posturing and ceremonial dancing before mating. Nests are a large mass of floating weed usually concealed in reeds or other lakeside vegetation. Here three to five eggs are laid. These hatch after four weeks and the young birds stay with the adults for a further four weeks. The boldly striped juveniles are frequently seen being carried on the backs of the parents from where they are taught to dive and feed.

Cormorant *Phalacrocorax carbo*

Common visitors along the coast, Cormorants are large (90cm) virtually all black birds, the only other similar bird being the Shag which is rarely encountered locally.

The head and body of the adults are glossy blue-black with white cheeks and throat, and during the breeding season they show white thigh patches. Juvenile birds are regularly seen; these are dark brown and have all white underparts.

Often seen fishing close inshore, they swim low in the water, their neck erect and bill held up at an angle. Skilled fishermen, they frequently dive for their exclusive diet of fish, bringing large prey to the surface where it is swallowed head-first.

Most numerous during the winter months, they can be seen swimming, fishing or loitering on wrecks, posts and sandbanks along the Swale. At Long Rock, flocks of over two hundred birds have been recorded on the sea during winter months.

When perched or resting they sometimes exhibit the habit of holding out their wings as if to dry them. The nearest breeding site is at Stodmarsh along the Stour Valley, where a small colony of thirty-two pairs raised sixty offspring during 1995. Nests are in trees at this site; in other areas they can be on cliffs or as at Dungeness, on the ground.

When breeding is over they disperse to find suitable overwintering sites around the British coastline and to various inland lakes.

Grey Heron *Ardea cinerea*

Unmistakably large birds (95cm), Grey Herons are regular visitors to the Whitstable coastline and can be seen throughout the year.

They can be seen feeding along the seashore at Seasalter and Long Rock, looking for small fish and molluscs. Preferring the dykes and water margins of Seasalter Marshes, their diet includes eels, frogs, fish and small mammals.

When fishing, they stand motionless waiting for prey to come within reach, then strike with their large dagger-like bill. Often cruising over suburban areas at dawn, they search for garden ponds, which unfortunately offer them a tasty breakfast especially during spring when they are feeding young.

Their distinctive slow and ponderous flight, trailing legs and a long retracted neck make the birds unmistakable. At dusk their movements are again noticeable, when they return to their evening roosts.

The nearest breeding site is at Graveney where a colony of about thirty pairs exist. The largest colony in England is at the R.S.P.B. Reserve at Northward Hill in Kent, where approximately two hundred pairs regularly breed. They nest close to one another at the tops of large trees, and use the same site every year. During February and March, three to five eggs are laid, which hatch after twenty-five days; the young are able to fly after seven weeks.

Brent Goose *Branta bernicla*

Brent Geese are common visitors and passage migrants and are some of the most obvious birds to be seen along the coast during autumn and winter. At high tide the birds often roost on the winter cereal crop behind the sea wall, much to the annoyance of local farmers. Long Rock has also become a regular wintering site and a flock of approximately three hundred birds has become usual.

Their main diet is eel grass *Zostera*, and other fine green seaweeds *Enteromorpha*, which grow abundantly on the Swale mudflats.

They are small geese (62cm) with dark heads, necks and bodies and have gleaming white undertails. Breeding success can be determined by carefully looking through the flocks. Young birds can be distinguished from adults by pale edging to their wing coverts and also the absence of the white neck patch. There are two races which are regularly found in the United Kingdom. These are the pale bellied *Branta bernicla hrota* which breeds in Greenland and Spitzbergen and winters in the North East of England and Ireland and the locally seen dark bellied goose *Branta bernicla bernicla* which breeds in the USSR and winters around the coasts of South East of England and Northern France.

The first migrants arrive at the end September, their numbers increasing to between one and two thousand during the next few months. They depart to their arctic breeding grounds during February and March to nest on the tundra.

Shelduck *top* Wigeon *bottom*

12

Shelduck
Tadorna tadorna

Shelduck are common winter visitors to the Swale Estuary, with some birds remaining throughout the summer to breed. Large ducks (60cm), they are conspicuous in their striking plumage; a white and black body with a large reddish-brown band just below the breast and a dark green glossy head. Both males and females have bright red bills, with the males having a large swelling at the base of the bill.

Generally seen feeding on the mussel banks along the Swale Estuary, flocks of up to two hundred and fifty birds are not unusual during the winter months. They can sometimes be seen sitting on the water in small groups at high tide.

Their diet consists mainly of small shellfish, a favourite being the small detritus-feeding snail *Hydrobia*, which is only three millimetres long and occurs in vast quantities.

Most Shelduck move north to breed. However, a few pairs stay on, breeding mainly in the Faversham area, where they nest in old rabbit burrows, and other various concealed hollows. Breeding starts at the end of April when clutches of between seven and twelve eggs are laid. Hatching after four weeks, the ducklings are led away by the parents to nursery areas.

During July and August most of the North European Shelducks migrate to the Waddensee on the North German coast, where they remain to moult en-masse, returning to their wintering sites during October and November.

Wigeon
Anas penelope

Wigeon are winter visitors occurring in huge numbers along the Swale Estuary. These small dabbling ducks (46cm), favour marshland where they graze on grass and other vegetation.

Large rafts of these ducks can be seen resting on the water in the mouth of the Swale and at low tide feeding along the water's edge. The Isle of Sheppey is their main stronghold where several thousand regularly spend the winter, with the RSPB's reserve at Elmley holding the bulk of the population.

Males are easily identified by their grey body, black undertail, chestnut head, and creamy buff crown patch. Females have an overall rich chestnut coloration and both sexes have white underparts. In flight, the males can be picked out by their bold, white forewing patches. This feature separates them from many other species. Males have a characteristic loud whistle which is a useful clue to identification when birds are seen at a distance.

Occasionally, a few birds can be found on Seasalter marshes where they are seen close to the dykes and other wet areas. During March and April the ducks begin their migration north to their main breeding grounds which extend from Iceland and east across Russia to the Pacific coast. There is also a small breeding population in the North of England and Scotland.

Seven to nine eggs are normally laid and incubation lasts for three and a half weeks. The ducklings are independent after seven weeks.

Teal *Anas crecca*

As with most duck, the diminutive Teal (36cm) occur mainly as winter visitors. They are usually seen in small groups mixed with flocks of Wigeon on the mudflats and the sea. The salt-marsh at Castle Coot also provides feeding and shelter, and at low tide they occasionally feed on nearby mussel banks and along the water's edge.

Teal are also found further west along the coast at Oare Marsh and Ham Pits, where they take advantage of the cover and food provided by reed-fringed lakes and wet areas.

The Swale area has a wintering population of over four thousand birds with numbers peaking during January. However, at the end of April most have migrated north to their breeding grounds in Northern Europe. Breeding takes place close to moorland pools, lakes and marshes. Nests are built on the ground amongst thick cover where they lay between eight and eleven eggs.

A wary duck and easily disturbed, they will take off, springing into the air with tremendous acceleration, swooping and wheeling in wader-like formations. From a distance they appear all dark except for the males which have a bright yellow patch on the sides of their tails. Closer inspection reveals that males have a chestnut head and neck with a green band extending from the eye to the nape and a greyish body. Females are a uniform mottled brown, whilst both genders have a green wing flash.

Eider *Somateria mollissima*

Eiders occur along the Swale as passage migrants and winter visitors, but are increasingly found during the summer months. From year to year the flock sizes of these sea duck vary from as few as half a dozen to as many as seventy-five or more.

Non-breeding summer birds tend to favour the Sheppey side of the Swale where they stay to feed and complete their summer moult. Towards the end of summer, post-breeding flocks often visit the Swale estuary. In August 1996 over two hundred birds were recorded off Shellness. Partially migratory, North European populations make southerly movements into the North Sea when their breeding territories ice over, often reaching south to the coast of France.

The British races, which have colonised the coasts of Scotland, Northern Ireland and Northumbria, are mainly sedentary. Nests are built amongst rocks, turf slopes or vegetation and are lined with abundant down. Here four to six eggs are laid and incubated for four weeks, the young birds becoming independent after ten weeks. When fully grown (60cm) the males are unmistakable, being mostly white with black flanks, crown and belly, and green nape. Females are quite plain and have a mottled brown coloration.

During summer, males undergo their eclipse moult where they loose their vivid plumage, changing to an all over dark brown similar to the females. When seen in flight, they fly low and in single file, whilst on land their upright stance enables them to walk surprisingly well. On the sea they often exhibit one of their display manoeuvres, which involves sitting back and flapping their wings, a good identification characteristic. Regularly diving to the seabed, they search for their diet of mussels, crabs and other molluscs.

Moorhen *top* Mallard *bottom*

Moorhen
Gallinula chloropus

Successful water birds, found in most countries throughout the world, the resident Moorhen can be difficult to find along the Whitstable coastline. Occasionally found at Long Rock, when habitat conditions permit, (one pair breeding in 1993 when a temporary lagoon was formed in the Brook), most records refer to single birds seen on the upper Brook, where the water is less salty.

There are regular sightings of birds in the dykes and ditches around Seasalter Marshes; however, it is difficult to estimate the true population. Rarely seen at all during the summer months owing to plant growth in the dykes, their presence is betrayed only by their characteristic 'krk' call from within the vegetation.

Usually secretive birds, they swim or run for cover when disturbed; however, they can be comparatively tame in city parks and ponds. Adults (33cm), are recognised by their all blackish plumage, small white wing patch and white undertail. The yellow tipped bill and red frontal shield are a striking feature, which make this species unlikely to be confused with any other.

There is some migration of Moorhens from Scandinavia and the Low Countries during winter. They are generally reluctant to fly and do so only at night, although if startled during the day will make low, short flights to the nearest cover. Their diet consists of mainly vegetable matter, seeds, fruit and insects. The bulky nest is made of dead plant material and situated amongst waterside vegetation where five to ten eggs are laid.

Mallard
Anas platyrhynchos

One of the most common and widespread ducks, Mallards are well represented along our coast throughout the year. They are partial migrants to our coastline during winter and in past years flocks of up to one hundred have gathered offshore at Seasalter, with many others found inland on the marshes.

Long Rock is another site which regularly supports a few Mallard, usually concealing themselves amongst vegetation on the salt-marsh. During spring, many birds stay on to breed using the dykes and reedbeds at Seasalter. Hidden amongst the vegetation, the nest is usually a shallow scrape on the ground, lined with down, grass and leaves. A large clutch of eggs is laid, usually between nine and thirteen, incubation taking about four weeks. The young fledge after a further eight weeks.

As with most surface feeding or dabbling ducks, females have drab brown coloration and are not easily recognized. During the breeding season the familiar male is readily identified by his bright yellow bill, dark green head, chestnut brown breast and grey body.

In June the males leave their young families and move off to moulting grounds, where they change into a brown eclipse plumage. This coloration gives them protective camouflage whilst they undertake their main moult, which is a period when they are flightless.

Red-breasted Merganser *Mergus serrator*

A member of the Sawbill family, the colourful Red-breasted Mergansers are regular winter visitors to the Swale Estuary and more commonly to Long Rock. It is not unusual here to see a dozen birds swimming and displaying to one another a few hundred metres offshore. Busy and active birds, they are expert "fishermen" and repeatedly dive to catch small fish and molluscs, never seeming to rest. In flight, their long outstretched necks and prominent large white wing patches help to confirm identification.

Males are conspicuous by their long slender red bills, dark green head and large wispy crest. Their bodies are mainly black and white with grey flanks, and a black-spotted, rusty brown breast. The female is dark grey with a small white speculum patch on the wing. The head and neck are brown with a less obvious crest.

They usually arrive in October and stay until April when they depart to their breeding grounds in Scandinavia and Northern Russia. A significant population breed in Scotland and Northern Ireland where they nest alongside sea-lochs, forested lakes and rivers.

Hen Harrier *Circus cyaneus*

The Hen Harrier is a large bird of prey (43-51cm) and is a frequent visitor to Seasalter and Nagden Marshes during the winter months where it is usually seen patrolling low over the fields and dykes looking for prey.

Holding its wings in a shallow 'V' it hunts with a wavering flight searching the rough ground for small mammals, and birds. Most of the birds come from the Isle of Sheppey, where they roost communally on the open fields and marshes. Requiring large tracks of open land, they are rarely seen in other parts of Whitstable, except perhaps flying over during migration.

The dark brown female, is identified by the white ring on the upper tail. The rarer male is quite different, slightly smaller, pale grey and has black wing tips. They can often be confused with the similar, but heavier built Marsh Harrier, which is mainly a summer breeding visitor to the Isle of Sheppey, and is sometimes seen hunting over the Seasalter marshes.

At the end of April, most Hen Harriers have departed to their moorland breeding grounds in the north of England and Scotland with others moving into Northern Europe. The nest is sited on the ground, amongst low cover where four to six eggs are laid. Incubation takes four weeks and the young birds fledge after a further five weeks.

Kestrel *Falco tinnunculus*

The only small resident falcon, Kestrels are familiar birds of prey often seen perched on roadside posts, or hovering over motorway verges. This new available habitat has benefited Kestrels, by providing large grassy tracts, where prey items have been able to thrive.

Along the coast they are not quite so common. Long Rock and Tankerton Slopes have rough grassy areas and the occasional Kestrel is sometimes seen hovering overhead, searching for voles and insects in the long grass. Seasalter also provides good habitat and Kestrels are regularly seen here throughout the year.

They maybe confused with the much rarer Sparrowhawk, which has rounder wing tips and a typical 'flap, flap, glide' flight action. Male Kestrels have reddish brown upperparts, pointed wings, blue-grey head and long tails with a black terminal band. Females are a more uniform reddish brown with black barring on the upperparts and tail.

Kestrels are the only birds of prey that we are likely to see hovering. By fanning their tail which is depressed downwards and flapping their wings, they are able to remain stationary in the air whilst scanning the ground below for food items.

The nest sites are varied and include cliff ledges, old crows nests, nest-boxes and locally on the ironwork of pylons at Nagden Marshes. Little or no nesting material is used and four to six eggs are normally laid during mid-April.

Oystercatcher *Haematopus ostralegus*

Probably the commonest wader to be seen along our shore, Oystercatchers are most abundant during the winter, although they can be seen in large numbers throughout the year.

The Swale supports a substantial population, which feed at low tide along the water's edge and on the mussel banks. Numbers range from a couple of hundred in late spring and early summer to perhaps five thousand during December and January.

Large black and white birds (43cm) with long red bills, Oystercatchers cannot be confused with any other shorebird. Probably the noisiest birds on the shoreline, their piping 'kleep, kleep' calls never seem to stop.

Regularly seen flying along the shore in small flocks, they twice daily commute from their high tide roost at Shellness beach and Harty Ferry to feeding grounds along the North Kent coast. Occasionally a few birds use the beach at Castle Coot as a roost site together with other wader species.

Their diet consists of cockles, mussels, worms and crabs. There have been attempts at breeding on Castle Coot and inland fields but most nests are deserted or predated. On the damp meadows of the Isle of Sheppey over one hundred and twenty pairs nested in 1995. However, only a small number actually fledged. The nest is a simple scrape on the ground where three eggs are laid. After hatching, the chicks remain with the parents for a further six weeks, learning various feeding techniques.

Ringed Plover *Charadrius hiaticula*

Present throughout the year, Ringed Plovers are mainly passage migrants and winter visitors. They are small waders (18cm) and frequently mix with Turnstones and Dunlin while roosting on the beach, these winter roosts often attracting over one hundred birds. They are identified by their grey back, white underparts, black neck band and black-tipped orange bill.

Feeding individually, they are not seen on the mudflats in large flocks. Preferring stony ground they are usually encountered along be-aches where their coloration gives superb camouflage, their presence given away only by a plaintive liquid call. This distress call is rarely noticed by beach walkers and holiday-makers, whose unintentional disturbance often causes nests to be deserted during the breeding season.

Most years there are about half a dozen pairs which attempt to breed along the coast, the majority at Seasalter. Surprisingly, amidst all the disturbance at Long Rock, a pair generally succeed in raising a brood most years. In 1997 two pairs successfully raised young on the beaches just west of the Sportsman Inn.

The nest, built on the upper beach, is a shallow scrape lined with a few specially selected stones or shells. Usually four eggs are laid at the beginning of April, often followed by another brood later on if conditions permit. Due to their incredible camouflage the eggs, nest, and young are virtually invisible, even when you are standing next to them. The eggs hatch after three to four weeks and the tiny chicks are able to run and follow their parents within hours of hatching.

Golden Plover *Pluvialis apricaria*

Golden Plovers are common winter visitors and passage migrants, but can be found at most times throughout the year except during late May, June and July when they are on their northerly breeding grounds.

Although they are waders, they favour coastal pasture and arable fields rather than the sandy shores and estuaries preferred by most other wading birds. Large flocks of five hundred or more birds can regularly be found on fields at Seasalter. Some also use the exposed mussel banks near Castle Coot on which to roost. Here their camouflage makes them difficult to detect and careful observation is required to locate them.

Rarely found singly, they flock together, unlike other members of the plover family. When seen in flight they often perform spectacular aerial manoeuvres just before landing.

In breeding plumage they can be confused with the silvery, Grey Plover, but the smaller Golden Plover (27-28cm), is speckled with gold and black on its upperparts. Both have black bellies, chest and face, but the Golden Plover shows a long white border between its upper and lower parts. Along with the Grey Plover they loose this black belly during the winter months and when in flight show an almost pure white underwing. They breed on upland moors and bogs in the north of England, Wales and Scotland where they nest on the ground, laying three to four eggs. They feed on beetles, worms and various insects employing the typical plover method of "snatch and run".

Grey Plover *Pluvialis squatarola*

Grey Plovers are common waders found along our shore from late July until late May the following year, confining themselves mainly to the mudflats of the Swale estuary. The stony flats at Long Rock also attract a small number of birds seen searching for their diet of worms, crustaceans and molluscs. When feeding, all plovers exhibit the typical gait of a few steps forward, followed by a pause to pick up food and then swallow.

Often seen standing motionless and looking dejected the Grey Plover completes this impression by uttering a mournful 'pee-ooee' whistle. Medium sized waders (29cm), they often appear dumpy with a hunched posture. In breeding plumage they are strikingly handsome birds with solid black underparts from belly to the bill and cheeks; their upperparts are silver-grey with black speckling. Non-breeding or winter plumage is quite different with largely white underparts and white fringed brown-grey uppers. In flight they are readily identified by the black inner underwing, a feature that the similar Golden Plover does not share.

At high tide, most roost with other waders on the Isle of Sheppey, a few staying on the beaches at Castle Coot and Long Rock.

Their arctic breeding range is almost circumpolar, extending from Northern Canada and Alaska, along to the Northern Russian tundra excluding Scandinavia, Iceland and Greenland.

They choose slightly higher ground for their simple nest and they lay their eggs at the end of June, the young birds fledging during August in time for the return migration.

Lapwing *Vanellus vanellus*

Lapwings are regular migrants and common winter visitors, often seen in large numbers along our coast and on farmland throughout the county.

They begin to arrive during July, numbers increasing throughout autumn and winter, especially if there are extremes of cold weather on the continent. Similarly, if we experience long spells of freezing temperatures when th-ey cannot feed, they will move off to warmer conditions.

Preferring open spaces, they can be found feeding on pasture and agricultural land where their diet consists of earthworms, leather-jackets, larvae and other insects. At 30cm, Lapwings are large plovers with a black neckband, white underparts, chestnut undertail and bronzy-green upperparts. Their most distinctive feature is the black crown and long crest. In flight, they are identified by their rounded black and white wings which slowly propel them along in ragged formations.

Returning to their breeding grounds during February and March they begin their aerial acrobatic courtship displays. A couple of pairs have bred on fields behind the Sportsman Public House in past years, but recently conditions have not been so favourable. Normally three to four eggs are laid at the end of March, using a crude scrape on the ground for a nest. The eggs hatch after approximately four weeks and the chicks leave the nest almost immediately, crouching down in suitable cover to avoid detection. The youngsters are able to fly and fend for themselves after five weeks.

Knot *Calidris canutus*

A small dumpy wader (25 cm.), the migrant Knot arrives on our shores in huge numbers during winter months. The first few passage migrants, still in their red summer plumage, are seen during August, stopping briefly to feed before continuing to wintering grounds as far south as Spain and North West Africa.

Large flocks arrive in November and December to spend the winter months feeding on the mudflats at Seasalter. Flock sizes vary from one to three thousand, but extreme cold weather will often attract double these numbers. In their pale grey winter plumage they are often confused with the smaller Dunlin. However, Knot are bigger, more rounded and have proportionately shorter legs and bill. Knot tend to feed together in a large compact flock, whereas Dunlin are inclined to spread out more thinly. They feed at a slower rate than Dunlin, the diet of both birds consisting of molluscs and crustaceans. When disturbed, the flocks perform spectacular aerial displays, wheeling above the mudflats in dense masses with continually changing formations. At the end of February most birds will have departed to the high Arctic with only a few passage birds seen in spring. At the end of June when the snow has receded, they lay four eggs in a shallow scrape on the open tundra. These hatch after three weeks and the young are led away and fed on insects, larvae and seeds. At this time, male birds take on the major role of rearing and protecting the young family.

Dunlin *Calidris alpina*

Passage migrants and winter visitors, Dunlin are one of the smallest (18cm.), and most common waders to be found along the seashore during these periods. In breeding plumage they can be identified alone by their black belly which remains until September when they moult into grey and white winter colours.

Appearing rather hunched, they feed actively, probing and pecking the mud with longish down-curved bills, searching for small molluscs and crustaceans, one favourite being the tiny snail *Hydrobia*, which occurs in enormous numbers.

At high tide most birds fly across to the huge wader roost on the salt-marshes near Harty Ferry, but some remain on this side of the Swale using some of the quieter beaches to rest. The beach to the east of the NRA Pumping Station at Seasalter and the shingle spit at Long Rock usually hold good numbers of Dunlin, Ringed Plover and Turnstone during winter high tides.

At low tide, mudflats at Seasalter support the major population of between one and three thousand birds. The stony shore at Tankerton and Long Rock offer less attractive feeding and support relatively fewer birds.

As with many waders, they begin to move north to their breeding grounds at the end of February followed by other passage birds until mid-May. They breed on Arctic tundra, wet moorland and montane habitats, laying four eggs in a scrape amongst low vegetation. The British Isles supports a population of 10,000 pairs, which nest mainly on the moors of Northern England and Scotland.

Black-tailed Godwit *Limosa limosa*

Elegant and long-legged, the Black-tailed Godwits are one of the largest waders (41 cm) to be found on the mudflats. Large numbers can be seen in late summer on the return migration, still in their breeding plumage. The reddish breast and neck gradually gives way to greyish tones as winter approaches. In flight they are readily identified by the broad white wing bar, black and white tail, long bill and trailing legs.

A regular place to find them is one mile west of the Sportsman public house at Seasalter, where they feed on the mussel banks. Sometimes large flocks can be encountered, four hundred and four being recorded one evening in July 1996. During late August, flocks begin to disperse, with a few individuals remaining into October and November.

There are wintering populations in other parts of Kent but the bulk of the birds move south into Africa. During early spring a few pairs return to breed on the North Kent Marshes but the majority breed in north central Europe and across through Asia with another population in Iceland.

Their preferred breeding habitat is marshland, damp meadows and fens. The nest is a substantial grass lined hollow amongst lush vegetation. Usually four eggs are laid at the beginning of May. Incubation takes twenty-four days after which the young leave the nest and are able to fly after four weeks. They feed on a variety of invertebrates and their long bills are suitably adapted to probe deep into soft soil and mud. The sensitive tip is able to seek out various prey items, which might include earthworms, molluscs and insects.

Bar-tailed Godwit *Limosa lapponica*

Bar-tailed Godwits are winter visitors and passage migrants. They arrive in August from their Arctic breeding grounds, still resplendent in their striking red-chestnut summer plumage, which extends down from the neck and breast to the undertail. The similar and slightly larger, Black-tailed Godwit shows this same red-chestnut coloration on the neck and breast, but it does not extend along to the undertail.

Bar-tailed Godwits (36-40cm), are identified in flight by their long bill, greyish-brown upperparts, and no wing bar, white 'V' on the back and grey barred tail. They are nearly always found on the mudflats between Seasalter Yacht Club and Blue Anchor P.H. Counts of between seventy-five and one hundred and fifty are regular in this area, whilst other parts of the coastline, particularly towards the east, attract relatively few.

When feeding, they spread out across the mudflats, walking briskly and probing with their very long bills, searching for molluscs and worms. Preferring firm ground on which to feed they generally stay away from the tideline, unlike the Black-tailed Godwits, which readily wade into the water. At the end of February most of the overwintering flock will move north to breeding grounds along the Arctic Circle, extending eastwards to the pacific coast of Siberia. These are later followed by passage migrants which follow through until May.

Whimbrel *Numenius phaeopus*

Passage migrants, Whimbrels are regularly seen along the coast from April to early May and again from late July until September during their return migration. Usually found on the mudflats or on the pastures at Seasalter, they rest and feed for a few days before continuing their journey. They are rarely found at Long Rock.

Looking very much like the Curlew, they can prove difficult to distinguish. Slightly smaller (40 cm), they have a shorter and straighter bill which droops at the tip and a dark crown with a central creamy stripe. All other plumage is similar to the Curlew. Usually their presence is first revealed by their characteristic call of seven shrill whistles, often made by birds in flight.

Along the coast they are more confiding than the curlew and regularly feed close to the beach. They rarely occur in large numbers, typical counts are between ten and twenty birds. They have a very northerly breeding distribution which includes Iceland, Scandinavia, Russia and Alaska. Our nearest breeding colony is on the Shetland Isles where there are approximately five hundred pairs.

Nesting on the ground in moorland and mountainous districts, they lay four eggs and incubation lasts nearly four weeks. Outside the breeding season they are birds of estuaries and rocky shores, where they probe for worms and molluscs, but will also feed on insects and larvae amongst rocks and seaweed.

Curlew *Numenius arquata*

The largest (50-60cm), and most widespread waders seen and heard along the coastline, Curlews are easily recognised by their size, long curved bills and overall brown-grey coloration. Very vocal, their melancholy whistles, "cour-lee", are commonly heard along the Swale.

Found in most months of the year they are generally considered passage migrants and winter visitors. During the early summer months, a few non-breeding birds can be found on the sandbanks along the Swale. At the end of July, passage migrants begin to arrive followed by wintering flocks, sometimes numbering five hundred birds.

High tide roosts in the fields behind Seasalter Yacht Club often contain one hundred and fifty birds, with other smaller roosts on fields next to the Swalecliffe Caravan Park and on Seasalter Levels, opposite Lucerne Drive.

They feed at a leisurely pace, probing the mud with their long bills, searching for ragworms and small molluscs. The length of the male's bill is 10-12cm and the female's even longer at 15cm. Most birds are found in muddy areas of the Swale but single birds can be found all along the coast. They return to their breeding grounds during February and March.

Many birds nest on the moorlands of Scotland, Wales and Northern England with scattered sites throughout the south-east and south-west. One clutch of four eggs is laid between April and June.

Redshank *Tringa totanus*

Redshanks are perhaps one of the most common and easily recognised waders along our coast. They are medium sized (27-29cm) and their long red legs separate them from nearly all other shore birds. In flight the broad white bar on the trailing edge of the wings and white rump can identify them. At rest, they have brown-grey upperparts and longish straight bills, which are red at the base. They are persistently noisy when alarmed and take off with a fast erratic flight, alerting all other birds in the area.

Mainly passage migrants and winter visitors, they can also be seen in small numbers during the breeding season when they nest along the Swale estuary. In common with other waders, they begin to arrive at the end of July, numbers at Seasalter often reaching four hundred or more during the winter. They prefer to feed on the open mudflats, searching for tiny crustaceans, mainly *Corophium*, which they consume at a phenomenal rate; some studies indicating forty thousand prey items can be taken each day. Sometimes, at high tide, they can be seen feeding along the water's edge at Seasalter or roosting with other waders at Long Rock and Castle Coot.

They breed in many parts of Britain, building a nest in damp meadows, salt-marshes and grassy moors. Four eggs are laid and are concealed in vegetation. The Isle of Sheppey supports approximately three hundred pairs and Seasalter Marshes almost certainly supports a couple of pairs. Breeding distribution is throughout Europe and across central Asia across to the pacific coast.

Greenshank *Tringa nebularia*

Less common than most other waders, Greenshanks regularly stop over on our coast during July to September, whilst on their southerly migration.

They are similar in appearance to the Redshank but are slightly larger (30-33cm), with greenish legs, a longer, slightly upcurved bill, pale grey-brown upperparts, streaked neck, breast, and white underparts. In flight, the wings are dark grey with no wing bar but the white rump is conspicuous as it extends halfway up the back. The Greenshank often utters a diagnostic 'chu-chu-chu' flight call when alarmed. They are found mainly at Seasalter, where they like to search for small crustaceans in the shallow pools on the musselbanks. Here they exhibit a very different feeding technique, which involves running and dashing across the shallows, chasing small fish fry and other prey items.

At high tide many birds roost on the pits and marshes at Faversham. However, none seem to favour any sites at Whitstable. The number of birds found off Whitstable is never great, varying from half a dozen to as many as fifty or sixty. Wintering further south, nearly all birds will have disappeared by the end of September seeking sites around the Mediterranean and on into Africa. Returning to their northerly breeding grounds, only a few birds are noted on spring migration. Most fly on to the marshes and swamps of Scandinavia and Central Asia. A small breeding population also exists in Scotland. The nest is a simple scrape on the ground and usually contains four eggs, which are laid in May or June.

Turnstone *Arenaria interpres*

Amongst the commonest waders along the shoreline, Turnstones are small (23cm), dumpy birds with a striking black, white and brown variegated plumage. Often seen feeding along the strandline at high tide, they busily pick through seaweed and flotsam using their short bills, looking for sand hoppers, small molluscs and seeds whereas, at low tide, they feed on the mussel banks and stony, 'hard' areas.

As with nearly all waders, they are present throughout the year except during the breeding season, but occasionally non-breeding birds can be found during the summer months. Large winter flocks can be encountered on the beaches with high tide roosts at Long Rock and Seasalter often exceeding four to five hundred birds. During peak migration, counts of over seven hundred have been recorded at Long Rock. At these roosts they will associate closely with Ringed Plover and Dunlin. Many Turnstones undertake long migrations and have one of the most northerly breeding ranges, with a complete circumpolar distribution. North American birds migrate as far south as South America and some Eastern Siberian birds winter in Australia and New Zealand. Turnstones nest on rocky islands and shingle beaches. Three to five light green eggs are laid between mid-May and July. Incubation takes three weeks, and the young birds fledge after a further four.

Black-headed Gull *Larus ridibundus*

Seen throughout the year, the resident Black-headed Gulls are common birds along the coast, breeding at several sites throughout Kent and the British Isles. During autumn and winter months, the population greatly increases with the arrival of passage and migrant birds from the continent.

During late afternoon in the winter, huge numbers of gulls fly in to roost on the Swale, several thousand birds being quite regular and at times the whole area looking white with gulls. In spring and summer they can be identified by their chocolate brown (not black) head, reddish-brown legs and bill. In flight they are recognised by a white wedge on the outer forewing. After breeding, adults moult into winter plumage, the dark brown cap changing to white except for a small dark patch just behind the eye. Juvenile and first winter birds show a brown coloration on their back, wings and head, the white tail having a pronounced black band. The legs and bill are pale carrot colour.

Black-headed Gulls are a widely distributed species breeding from Iceland, Europe and across through central Asia to the Pacific coast. They nest in colonies and the typical habitats include gravel pits, dunes and coastal marshes. Locally they nest at Windmill Creek on the Isle of Sheppey and on islands in the Medway estuary. Usually three eggs are laid between May and June. Incubation lasts three to four weeks, and the young fledge after a further five to six weeks.

Common Gull *Larus canus*

Virtually absent during the breeding season, Common Gulls are mainly passage migrants and winter visitors, though not always as common as the name suggests. Similar in size, (43cm), to the Black-headed Gull they look similar to the larger Herring Gull, but have greenish-yellow legs and bills, darker grey mantle and daintier appearance. Immature birds have dark wing tips and trailing edges, with a black terminal band on the tail; the head and neck are streaked. As with all gulls, remnants of immature plumage can remain for a long time. Full adult plumage for the Common Gull is not attained until the third winter. They are found in various habitats, ranging from agricultural land, town parks, playing fields and coastal mudflats. However, they never venture far out to sea. A regular place to find them is the NRA sluice at Seasalter.

Their varied diet includes worms, molluscs, crustaceans, and any other morsels that might be scavenged. Breeding distribution extends east from northern Britain, Scandinavia, the Low Countries, central Asia, across to the Pacific coast and over to Alaska and western Canada. They nest on the ground in small colonies beside lochs, on shingle banks and on moorland. The nest is lined with dry grass, heather or other plant material. One brood of three eggs is laid between the end of April and the beginning of June.

Herring Gull *Larus argentatus*

Large and heavily built (53-59cm), Herring Gulls are winter visitors and passage migrants with an increasing resident breeding population.

Similar to the smaller Common Gull they also have a pale grey back but with a heavier, bright yellow bill with a red spot near its tip, fierce yellow eyes and greyish-pink legs. During their first year, juvenile birds have all black bills and appear dark streaked, changing the following summer to a more overall dirty grey appearance. Second winter/summer plumage gradually assumes the grey back of the adult bird, a white tail having a dark subterminal band. After four years full adult plumage is achieved. Scavengers, Herring Gulls eat almost anything, feeding at rubbish tips, harbours, fields and along the coast, they take a variety of food including shellfish, plant material, birds, worms and offal.

Nesting starts in April, although birds will return to their chosen breeding sites several weeks before, where they sit and wait, guarding their territories against possible interlopers. They usually nest in colonies. The cliffs at Dover support a population of approximately sixty pairs which is easily viewed from the top of Langdon Cliffs. Locally, they nest on factory roofs at the two industrial estates along the Thanet Way. The nest is a large mass of vegetation where two or three eggs are laid. These hatch after four weeks, both parents incubating the eggs. Young birds stay at the nest site for a further two months before becoming independent.

Great Black-backed Gull *Larus marinus*

Largest of all the gulls, Great Black-backed Gulls (64-78cm) have a wingspan of nearly five and a half feet and can only be confused with the less common and smaller Lesser Black-backed Gulls (52-67cm), which have yellow legs and slate grey backs.

Generally, they are winter visitors and passage migrants but normally there are a few non-breeding birds loitering offshore throughout the summer. A regular place to find them during winter is on the raised 'shingle island', just off shore at Seasalter, opposite the beach chalets. Either side of high tide, many different gulls can be seen resting here, giving the observer a good opportunity to work out the identification of the five different gull species that regularly occur.

Found all along the coast and roosting on fields, they can be identified by their large size and black backs, although immature plumages and occasional Lesser Black-backed gulls might cause confusion. Adults have a large, yellow bill, which has helped them to become a major predator of birds and mammals during the breeding season. Raiding the nests of other birds, they will swallow small chicks whole, and even take adult birds. Outside of the breeding season they exist by scavenging and taking anything that might be considered edible.

Sandwich Tern *Sterna sandvicensis*

First described in 1787 from birds observed in Sandwich, Kent, these large terns (36-41cm) are regular passage migrants and summer visitors. Returning from the west coast of Africa in March, they are one of the earliest migrating birds to return to our coastline, and breed in scattered colonies, mainly along the east and south coasts.

They are the largest terns to visit our shore. Adult birds are identified by their very long thin wings, short forked tail and long slender black bill with yellow tip. The head has a particularly striking black shaggy cap.

They feed on sprats, sand eels and other small fish which are caught by diving from heights of about 7 metres above the water, often hovering before making their plunge. They are normally seen flying along our coast in small numbers during spring and autumn, often stopping to rest on posts and breakwaters, giving a good opportunity to note their features. There are no local breeding sites. The closest is at the R.S.P.B. reserve at Dungeness, although in certain years, the birds prefer to move west to the reserve at Rye Harbour in East Sussex. At these reserves they nest in colonies on specially created shingle islands where two dark-blotched, sandy-brown eggs are laid, giving them excellent camouflage. Incubation lasts for just over three weeks, the young chicks fledging and able to fly after a further five weeks.

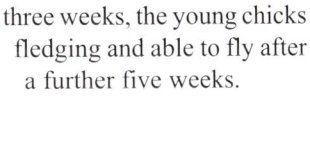

Common Tern *top* Little Tern *bottom*

Common Tern
Sterna hirundo

Common Terns are passage migrants and summer visitors, breeding in small numbers on the Isle of Sheppey. First migrants arrive during April when they can be spotted flying along the shore, often diving into the sea for small fish. During the breeding season it is usual to find half a dozen birds patrolling along the Swale searching for food, frequently hovering and taking aim before diving down onto their intended prey.

The large dykes on Seasalter marshes also provide good hunting grounds for terns, fish fry and insects being the main attraction. Spending most of their life on the wing, they have a buoyant flight action with strong deliberate wing beats. Medium sized birds (31-35cm), they have glossy black caps and black tipped red bills, pale grey wings and very short red legs.

Common Terns breed throughout most of Britain; our nearest colony is on the Isle of Sheppey, where in 1995 twenty pairs nested at Windmill Creek. There are attempts at breeding on the beach at Castle Coot, near Faversham Creek but these nearly always result in desertion or predation. The nest is a shallow scrape amongst the stones with little or no lining. Three eggs are laid which hatch after three weeks, and the young birds are able to fly after a further four weeks.

Little Tern
Sterna albifrons

Smallest of the sea terns (23-26cm), Little Terns are passage migrants and summer visitors, which spend the winter months off the coast of Africa along with other tern species. The small sized, squat bodies, yellow legs and black tipped yellow bills, coupled with very fast wing beats readily identify Little Terns at a considerable distance. Closer inspection reveals a black cap with a white forehead, pale grey wings with dark leading edges and yellow legs.

Arriving in April they are often seen at Long Rock or further west in the mouth of the Swale. Numbers are never high, but post migration flocks of nearly forty birds, remained in the Swale during August 1997. At this time they can be seen flying up and down the estuary in loose formations frequently diving and rapidly hovering before taking the plunge. Nationally rare breeding birds (less than two and a half thousand pairs), the south-east coast supports a large percentage. The Swale and Medway estuaries generally have several breeding pairs whose success varies from year to year. Their preference to nest on sand and shingle beaches makes them particularly vulnerable to human disturbance along with tidal flooding and predation. Most years, Castle Coot attracts two or three pairs although productivity is often disappointing. Shellness and Windmill Creek also offer suitable habitat. Long Rock would be an ideal breeding site for these dainty birds but human pressures have now made this impossible.

Short-eared Owl *Asio flammeus*

Widely distributed throughout the Northern Hemisphere in both the New and Old Worlds, Short-eared Owls have become a flourishing species with other populations inhabiting much of South America and Pacific islands. The British population of breeding Short-eared Owls is mainly confined to the northern half of the country with a few isolated pairs on the salt-marshes around the south-east coast.

The bulk of the Kent records concern wintering birds and passage migrants. However, in some years, there are a few individuals which breed on the saltings of the Medway estuary and the Isle of Sheppey. Numbers of birds fluctuate greatly according to the population cycle of the short-tailed field vole, which forms a major part of their diet. These large owls (38cm) are easily spotted quartering fields and marshes, their long wings often held in a shallow 'V', whilst patrolling the area with a slow, wavering, jerky flight. Upperparts are tawny buff whilst the underwings are white and marked by dark carpal patches and have barring on the outer tips. They have a large rounded face with two fierce, dark ringed, yellow eyes and their breast is heavily streaked. Locally, Seasalter is the only place to find this day flying owl, although in recent years they have become rather infrequent. Breeding is mainly confined to the large open tracts of moorland and young forestry plantations of upland Britain. Nests are made on the ground and concealed by vegetation. Their four to seven eggs hatch after four weeks, young owlets requiring another four weeks of care before they are able to fly.

Kingfisher *Alcedo atthis*

The brilliant electric blue flash of a passing Kingfisher makes it one of the most exciting and splendid birds to see along our coast. Sedentary and resident on many rivers and lakes, we generally only see them during the winter months, when the colder climes of northern Britain force many Kingfishers to move south, where they can feed along our relatively ice-free waters.

The dykes and tidal pools at Seasalter regularly attract wintering birds, which occasionally can be seen fishing from breakwaters at low tide or hovering over pools on Castle Coot. The Brook, at Long Rock is another reliable site to find these inconspicuous birds. They are generally seen only when they take flight, speeding low over the water, often accompanied by a high pitched whistling. Identified by their small size (18cm.) and enormous long bill, their brilliantly coloured blue and orange plumage makes this exotic species, unmistakable. Bill coloration separates the sexes. Males have an all black bill but the females have an orange base to their lower mandible. Breeding is unlikely in the Whitstable area, as nesting sites are usually in banks next to rivers or lakes where there is a good supply of small fish. They excavate a tunnel, one to three feet long and lay between six and seven eggs. These hatch after about three weeks. The diet consists mainly of small fish, tadpoles and insects, which are caught by plunge diving from an exposed perch overlooking the water.

Skylark *top* Linnet *bottom*

Skylark
Alauda arvensis

A nationally declining species, Skylarks can still be found in reasonable numbers along our coastline during the winter months. Many of these birds are migrants, escaping the cold weather of northern and eastern Europe and arriving here during October and November. Feeding on small seeds, plant material and insects, they can be found almost anywhere along the coast especially on rough grassland, beaches and salt-marshes.

Sometimes very confiding, Skylarks can be approached to within a few feet before flying away at the very last moment. When alarmed, they adopt an upright stance with raised crests. However, when feeding they often take on a very low creeping posture. Their small size (17-19cm) and streaky brown coloration makes them appear dull compared to other birds, but during courtship their remarkable song flight makes them unmistakable. Climbing up almost out of sight they soar, hanging in the sky for several minutes, pouring out their musical song before dropping to the ground. In flight they are recognisable by their short, blackish tails, broad wings which have white trailing edges, creamy-buff colour underparts and streaked breasts.

Skylarks are birds of open country and fields, breeding throughout Europe, Central Asia and across to the Pacific Coast. They nest on the ground in grassy tussocks, laying between three and five eggs, which hatch after eleven to fourteen days and the young birds are able to fly after three weeks.

Linnet
Carduelis cannabina

Common and widespread residents, Linnets are often found in large flocks along the coast during the autumn migration period, when they tend to desert their breeding habitats and move to coastal sites such as saltings, seashore, waste ground and stubble fields to feed. At this time they have rather drab plumage, streaky, reddish-brown backs and grey heads and in flight show some white in their wings and tail. During the spring the males becomes more colourful. The breast and crown change to bright red and the back to a rich chestnut, females remaining relatively unchanged.

Regular wintering flocks can be found on the beach just beyond the Sportsman P.H. Here, autumn flocks of between fifty and one hundred birds can often be seen feeding amongst vegetation on the upper beach, usually in the company of Greenfinches, Goldfinches, Meadow Pipits and Skylarks. Smaller flocks also feed on the shingle spit at Long Rock and individual birds can be seen almost anywhere. In flight they can be recognised by their chattering and twittering song and even when feeding often continue this relentless chatter.

Breeding starts in mid-April, and the scrubby cover on Duncan Downs supports many nesting birds as do hedgerows and gardens. The nest is constructed low down in a bush, normally no higher than five feet and is a bulky cup of grass and plant stems lined with fine grasses, hair and wool. Five to six eggs are normally laid and incubated entirely by the female.

Meadow Pipit *Anthus pratensis*

Similar in coloration to the Skylark, Meadow Pipits are smaller (14-15cm), and much slimmer birds. Their olive-brown upperparts are streaked brownish-black and the buff-coloured breast and flanks are heavily marked with dark spots. In low level flight the white outer tail feathers can be very conspicuous along with their long thin bill.

Meadow Pipits are very common and are found in grassy places; they can be seen all along the coast in suitable habitats, from Long Rock to Seasalter where they can be found throughout the year. A resident breeding species, the population swells during winter months when influxes of northern and continental migrants visit our shores to escape the harsh conditions of their breeding grounds. Some of these birds migrate further to southern Europe. They are usually first detected by their high pitched, 'seep, seep, seep', call when taking to the air in alarm, and during the breeding season. They can also be located by their series of accelerating 'zi zi zi zi' calls whilst rising into the air, after which they parachute down into the grass uttering a slower, 'seea seea seea seea' trill. Breeding distribution extends from Iceland, across northern Europe and into the north-western parts of Russia.

The nest is concealed in a grassy tussock, where four to five eggs are laid. The eggs are incubated for two weeks and the young chicks fledge two weeks after hatching. Their diet consists mainly of small insects, spiders and occasionally seeds.

Yellow Wagtail *Motacilla flava flavissima*

Summer breeding visitors, Yellow Wagtails (16.5cm) arrive during April and can be observed migrating along our coastline after wintering in West Africa. There are many races of yellow wagtail across Europe and Asia all exhibiting a variety of head patterns and colours. The race found in England is '*M. f. flav-issima*' but on occasions the blue-headed form, '*M. f. flava*' has been recorded along our coast.

Graceful and slim birds found typically in damp meadowland, they associate with cattle to take advantage of the disturbed insects upon which they feed. The males usually arrive two weeks before the females and have long tails with yellow underparts, greenish back and bright yellow foreheads. The females have browner upperparts and are paler yellow below. They are active birds, and are often seen walking, running, tail wagging and darting into the air to catch insects.

Their flight is very undulating and they often utter a high-pitched 'tsweep' call betraying their presence overhead. Locally, a few pairs may be encountered at Seasalter Marshes where breeding occurs between May and July. The nest is made on the ground and is lined with grass and hair. Usually five eggs are laid and the young birds fledge after four weeks. During August post migratory flocks of birds can be seen feeding on the seawall at Seasalter and gathering in the evening before roosting in the reedbeds. During September and October they depart for their winter quarters in West Africa.

Stonechat *Saxicola torquata*

The tiny Stonechats (12cm.) are some of the most delightful birds to visit our coast during the winter months. At this time of the year they have buff edgings to most of their feathers. This gradually wears off by the spring to reveal their striking summer plumage. In breeding plumage, the brightly coloured males have pale chestnut underparts, white neck patches, black heads and almost all black up-perparts, the wings showing small white panels. Females are less colourful, predominantly brown and chestnut and showing less white. Locally they are not common. Most winters only two or three pairs are found along our coast.

The most reliable sites are near the Sportsman public house, Long Rock and on the Seasalter Marshes, where they are usually seen perched on low vegetation and fence posts. Most birds spend the winter at coastal sites around the country and during early spring return to their upland or heathland breeding grounds. In the spring of 1996, a pair of Stonechats stayed on to breed at Seasalter, near the main drain pumping station.

Several other coastal sites in Kent also attract an occasional breeding pair, the total rarely exceeding ten pairs. On a global scale, Stonechats are a widespread breeding species ranging throughout Europe, most of Asia, across to Japan and through most of Africa, south to the Cape. The nest is a small grassy cup, sited on the ground often below gorse or bramble. Five eggs are laid and hatch after two weeks, the young are then fed on their staple diet of insects.

Wheatear *Oenanthe oenanthe*

Regular passage migrants, Wheatears arrive from their African winter quarters during April and May pausing to feed before continuing on to their northerly breeding grounds, returning again between August and October.

They are usually found at Long Rock and along the coastal fringe at Seasalter, particularly on the stretch of shingle and grass which extends from the beach huts by the Sportsman public house, westwards to the start of the sea wall.

They are small birds (15 cm.) and in the spring males are most handsome with their grey crown and back, black wings, black face mask, sandy-buff breast and white belly. Females are duller, with a dark brown mask and are more brown above. In flight, both sexes show a white rump, which is bordered by a black inverted 'T' band on the tail. They are often seen perched on the sea wall, standing very erect, making short low flights to other parts of the wall or down to the ground to catch insects. There are usually no more than half a dozen birds present at Seasalter and similar numbers at Long Rock, where they also favour the short grass of the new football pitch.

Their main breeding habitat is in the north and west of the U.K. where they concentrate in areas above three hundred metres. There are also scattered sites across lowland Britain, including several in Kent; Dungeness supported seventeen pairs in 1995. In recent years there have been records of young being fed by adults on the beach at Seasalter. Crevices amongst rocks and rabbit burrows are typical nest sites.

Reed Warbler *top* Sedge Warbler *bottom*

Sedge Warbler
Acrocephalus schoenobaenus

Often confused with Reed Warblers, Sedge Warblers are earlier migrants, arriving at the beginning of April and breeding in the same reedbed habitat. They are more widely distributed and breed throughout the British Isles, and on the continent, as far north as Finland.

Reed Warblers have a more southerly distribution, only breeding as far north as South Yorkshire in the UK. Sedge Warblers are distinguished from other reedbed warblers by their streaked upperparts, bold creamy eyestripe and dark crown. They skulk in the reedbeds, but frequently come up to an exposed perch to utter their scratchy song, which is more varied and less repetitive than the Reed Warbler's. They often sing from the tops of bushes, but if no suitable perch is available will often be seen performing an ascending song flight, then 'parachuting' back to the ground.

Due to their habitat requirements, the bulk of all records are from Seasalter marshes, with only occasional sightings from Long Rock. Nests are built close to the ground amidst thick vegetation and reeds; five eggs are laid towards the end of May. Incubation takes nearly two weeks and the young chicks are ready to leave the nest after a further ten to twelve days. Feeding on various insects, they nearly double their weight, and are then ready for their return migration to Africa during August and September.

Reed Warbler
Acrocephalus scirpaceus

Summer breeding visitors, the tiny Reed Warblers (13 cm), arrive from their African winter quarters during April and depart during August. They are numerous on the Seasalter Marshes, most reed-fringed dykes holding several pairs. In May 1996, a population survey on the Seasalter Levels Local Nature Reserve, revealed one hundred and one birds along the dykes of the nineteen fields (178 acres).

The dykes behind the sea wall, west of the Sportsman pub are perhaps one the best places to see these birds. There are few other places in Whitstable where they occur, Long Rock seldom producing many records. Usually secretive, they move about within the reedbed, their presence often betrayed by their chattering phrases. However, they occasionally show themselves at the top of the reeds. Very plain looking birds, they have a long bill, sloping forehead and a rounded tail. Their upperparts are rufous brown, underparts buff-white, flanks tinged brown and the rump a rusty brown. Both sexes are alike.

As with many migrant birds, males are the first to arrive back from their wintering grounds and on arrival start to stake out territories by singing on top of a convenient reed. The deep cupped nest is interwoven amongst four or five reed stems by the female and within one week the first of four eggs are laid. After eleven days the eggs hatch and chicks are fed on a diet of insects. After another two weeks the young birds are ready to leave the nest.

Bearded Reedling *Panurus biarmicus*

The strikingly attractive Bearded Reedlings have only been found along our coastline during the last decade. A rare and resident species, only a handful of birds were known to exist in the county before the Second World War, after which they slowly established themselves.

Confining themselves to life in the reedbeds, a small resident population now exists at the western end of Seasalter, towards Faversham Creek. During winter months single birds have been recorded feeding along the reed-lined dykes behind the N.R.A. pumping station at the eastern end of Seasalter. Usually difficult to see, their presence can often be detected by their characteristic "pinging" contact call. They are small birds (16.5cm), recognised by their mainly tawny brown coloration and a long tail, the males having long, black drooping moustaches, and grey heads. They are very active, climbing and hopping through the reedbeds and occasionally making short low flights over the vegetation, making them difficult to find.

They nest low down among the reed stems, laying between five and eight eggs. Incubation takes eleven to thirteen days and the young birds fledge after approximately two weeks. During the spring and summer months, the species is mainly insectivorous but during winter their diet consists mainly of seeds from Reedmace and Phragmites.

Snow Bunting *Plectrophenax nivalis*

Snow Buntings are small colourful birds (16.5cm), occasionally found along our coast during the winter months. Easily identified from other wintering passerines by their variegated white plumage, the male birds look almost pure white in flight, apart from their black wing tips.

Usually seen foraging along the beaches looking for seeds and insects, their numbers vary from single birds to small flocks of approximately fifty birds. Regularly seen at Long Rock and along the beach at Seasalter, they are well camouflaged amongst the shingle. When disturbed they will all lift off together, flying up and down the beach uttering their characteristic trilling calls before landing. These birds regularly fly across to Shellness beach on the south-eastern tip of the Isle of Sheppey to feed, especially if there has been too much disturbance on the Whitstable coast.

In favourable years the earliest birds arrive in October, followed by others which usually remain throughout the winter. It should be noted that it is not every year that we are fortunate enough to see these attractive birds. Leaving our coast in February they return to their main breeding grounds on the Arctic tundra and mountains of Scandinavia. There is also a small population of less than one hundred pairs breeding in Scotland. They nest in rocky crevices and on stony terrain laying four to eight eggs.

Reed Bunting *top* Corn Bunting *bottom*

Reed Bunting
Emberiza schoeniclus

Reed Buntings are common resident birds inhabiting many of the wet areas bordering our coastline. At Seasalter there is a healthy breeding population along the dyke and ditch systems, with Long Rock supporting a few pairs. They are generally only seen in wet marshy areas, but the gradual disappearance of this habitat has encouraged these birds to explore drier localities and it is not uncommon to see them feeding on stubble fields and along hedgerows during the winter.

The male's breeding plumage is unmistakable; striking black head and throat, white neck and moustachial stripe with a streaky chestnut coloration over the back, white breast and underparts. Females do not have the black head or white collar, and their underparts are more streaked. Both birds have white outer tail feathers, which are quite prominent in flight. Male birds are often seen on the top of reed stems uttering their "tsee-u " call note, which betrays their presence.

Nests are usually found close to the ground, in sedges, tussocks or low bushes where five eggs are laid at the end of April. These hatch after two weeks and a further ten to twelve days of development is required before the young birds are ready to leave the nest. Much of their food is found among the reedbeds and bushes and consists of insects and seeds. During the winter, many birds can be found feeding along the upper tideline, and reedbeds at Castle Coot and also on the beach at Long Rock.

Corn Bunting
Miliaria calandra

Corn buntings are gregarious winter visitors regularly found at Long Rock, usually concealed in long grass where they busily feed on seeds and insects. They are easily flushed by walkers along the path around the Brook and often fly up to the bushes nearby. Numbers can be considerable with recent counts of up to one hundred and fifty birds, but forty or fifty are more commonplace. The other place to look for them is at Seasalter where they search for seeds amongst the hay put out for the sheep. A patch behind the Yacht Club during the winter of 1997-1998 regularly attracted thirty birds, often showing themselves when they flew up to perch on telegraph wires. The majority disperse when the breeding season approaches, but one or two pairs remain to breed on the arable fields and marginal land further west beyond the Sportsman public house.

During the summer, their distinctive song said to resemble the jangling of a bunch of keys, can often be heard from the fields behind the seawall at Seasalter . If you are lucky, displaying birds might show their habit of flying with legs dangling. Typical brown birds, they are fairly large (18cm), thickset and have comparatively short tails. The top of the head appears rather flat and the bill is stubby and yellowish. Upperparts are brown streaked black, breast and belly are creamy-buff with fine dark streaking on the flanks which come together to form a darkish breast patch; the throat is pale buff. Nesting on the ground, they lay three to five eggs.

Checklist

The following birds have been recorded within the boundary of the old Whitstable Urban District Council and westwards, to Faversham Creek including those seen flying over the sea and up into the Swale. The boundary starts in the east at Long Rock, then tracks south on the eastern side of Chestfield, passing through a large section of Thornden and Clowes Woods, then cuts across to the south of Ellenden Wood and over to the south-western end of Yorkletts. From this point the boundary changes and continues along Monkshill Road, left at Faversham Road and finally turns right along Sandbanks Road to Nagden Cottages. These are records of wild birds, seen, or accepted by the contributors during the last decade.

Red-throated Diver *Gavia stellata*
  Regular winter visitor
Black-throated Diver *Gavia arctica*
  Rare winter visitor
Great Northern Diver *Gavia immer*
  Rare winter visitor
Little Grebe *Tachybaptus ruficollis*
  Scarce winter visitor
Great Crested Grebe *Podiceps cristatus*
  Common winter visitor
Red-necked Grebe *Podiceps grisegena*
  Scarce winter visitor
Slavonian Grebe *Podiceps auritus*
  Scarce winter visitor
Fulmar *Fulmarus glacialis*
  Irregular visitor
Manx Shearwater *Puffinus puffinus*
  Rare autumn passage migrant
Leach's Petrel *Oceanodroma leucorhoa*
  Rare autumn passage migrant
Gannet *Sula bassana*
  Irregular passage migrant
Cormorant *Phalocrocorax carbo*
  Common winter visitor, fewer in summer
Shag *Phalocrocorax aristotelis*
  Rare winter visitor
Cattle Egret *Bubulcus ibis*
  Very rare vagrant
Little Egret *Egretta garzetta*
  Becoming a regular visitor
Grey Heron *Ardea cinerea*
  Common visitor all year, mainly from Graveney breeding colony
White Stork *Ciconia ciconia*
  Rare visitor

Glossy Ibis *Plegadis falcinellus*
  Rare visitor
Spoonbill *Platalea leucorodia*
  Rare spring and summer visitor
Mute Swan *Cygnus olor*
  Breeding species, winter visitor
Bewick's Swan *Cygnus columbianus*
  Scarce winter visitor
Whooper Swan *Cygnus cygnus*
  Scarce winter visitor
Bean Goose *Anser fabalis*
  Rare winter visitor
Pink-footed Goose *Anser brachyrhynchus*
  Rare winter visitor
White-fronted Goose *Anser albifrons*
  Scarce winter visitor
Greylag Goose *Anser anser*
  Scarce visitor
Canada Goose *Branta canadensis*
  Scarce visitor
Brent Goose *Branta bernicla*
  Common winter visitor
Egyptian Goose *Alopochen aegyptiacus*
  Rare visitor
Shelduck *Tadorna tadorna*
  Common winter visitor with a small breeding population
Mandarin Duck *Aix galericulata*
  Rare visitor
Wigeon *Anas penelope*
  Common winter visitor
Gadwall *Anas strepera*
  Scarce winter visitor
Teal *Anas crecca*
  Common winter visitor

Mallard *Anas platyrhynchos*
   Common breeding resident
Pintail *Anas acuta*
   Scarce winter visitor
Garganey *Anas querquedula*
   Rare spring migrant
Shoveler *Anas clypeata*
   Occasional winter visitor
Red-crested Pochard *Netta rufina*
   Rare vagrant
Pochard *Aythya ferina*
   Scarce winter visitor
Tufted duck *Aythya fuligula*
   Occasional winter visitor
Scaup *Aythya marila*
   Regular winter visitor
Eider *Somateria mollissima*
   Common, all year
Long-tailed Duck *Clangula hyemalis*
   Rare winter visitor
Common Scoter *Melanitta nigra*
   Regular winter migrant
Velvet Scoter *Melanitta fusca*
   Regular winter visitor most years
Goldeneye *Bucephala clangula*
   Scarce, winter, usually in flight
Red-breasted Merganser *Mergus serrator*
   Common winter visitor
Goosander *Mergus merganser*
   Rare winter visitor, usually in flight
Common Buzzard *Buteo buteo*
   Rare, seen on passage.
Red Kite *Milvus milvus*
   Rare Vagrant
Marsh Harrier *Circus aeruginosus*
   Occasional summer visitor
Hen Harrier *Circus cyaneus*
   Winter visitor
Montagu's Harrier *Circus pygargus*
   Rare summer visitor, has attempted to breed
Sparrowhawk *Accipiter nisus*
   Breeding resident
Osprey *Pandion haliaetus*
   Scarce passage migrant
Kestrel *Falco tinnunculus*
   Common breeding resident
Merlin *Falco columbarius*
   Regular winter visitor
Hobby *Falco subbuteo*
   Regular summer visitor
Peregrine *Falco peregrinus*
   Regular winter visitor

Red-legged Partridge *Alectoris rufa*
   Common breeding resident
Grey Partridge *Perdix perdix*
   Breeding resident, declining rapidly
Quail *Coturnix coturnix*
   Rare passage migrant
Pheasant *Phasianus colchicus*
   Common breeding resident
Water Rail *Rallus aquaticus*
   Scarce winter visitor, possibly bred
Moorhen *Gallinula chloropus*
   Common breeding resident
Coot *Fulica atra*
   Summer breeding visitor
Crane *Grus grus*
   Very rare vagrant
Oystercatcher *Haemotopus ostralegus*
   Very common winter visitor, a few pairs stay to breed
Avocet *Recurvirostra avosetta*
   Summer visitor from Sheppey breeding colonies, some wintering birds
Little Ringed Plover *Charadrius dubius*
   Rare passage migrant
Ringed Plover *Charadrius hiaticula*
   Common winter visitor, breeds locally
Dotterel *Charadrius morinellus*
   Rare passage migrant
Golden Plover *Pluvialis apricaria*
   Common winter visitor, passage migrant
Grey Plover *Pluvialis squatarola*
   Common winter visitor, passage migrant
Lapwing *Vanellus vanellus*
   Common winter visitor, scarce breeding species
Knot *Calidris canutus*
   Common winter visitor
Sanderling *Calidris alba*
   Winter visitor to Long Rock, increasing
Little Stint *Caldris minuta*
   Rare passage migrant
Curlew Sandpiper *Calidris ferruginea*
   Rare passage migrant
Purple Sandpiper *Calidris maritima*
   A few regularly winter at Long Rock
Dunlin *Calidris alpina*
   Very common winter visitor, passage migrant
Ruff *Philomachus pugnax*
   Scarce winter vistor/migrant
Jack Snipe *Lymnocryptes minimus*
   Occasional winter visitor

Snipe *Gallinago gallinago*
  Common winter visitor
Woodcock *Scolopax rusticola*
  Scarce woodland resident, possibly breeding
Black-tailed Godwit *Limosa limosa*
  Common passage migrant, a few occasionally overwinter
Bar-tailed Godwit *Limosa lapponica*
  Common winter visitor, passage migrant
Whimbrel *Numenius phaeopus*
  Regular passage migrant
Curlew *Numenius arquata*
  Common winter visitor
Spotted Redshank *Tringa erythropus*
  Scarce passage migrant
Redshank *Tringa totanus*
  Common winter visitor, passage migrant, a few pairs breed
Greenshank *Tringa nebularia*
  Fairly common passage migrant
Green Sandpiper *Tringa ochropus*
  Occasional winter visitor and passage migrant
Common Sandpiper *Actitis hypoleucos*
  Regular passage migrant
Turnstone *Arenaria interpres*
  Common winter visitor/passage migrant
Grey Phalarope *Phalaropus fulicarius*
  Rare migrant
Pomarine Skua *Stercorarius pomarinus*
  Occasional autumn passage migrant
Arctic Skua *Stercorarius parasiticus*
  Regular autumn passage migrant
Long-tailed Skua *Stercorarius longicaudus*
  Scarce autumn passage migrant
Great Skua *Stercorarius skua*
  Regular autumn passage migrant
Mediterranean Gull *Larus melanocephalus*
  Regular overwintering bird at Long Rock
Little Gull *Larus minutus*
  Occasional passage migrant
Sabine's Gull *Larus sabini*
  Rare autumn migrant
Black-headed Gull *Larus ridibundus*
  Common all year
Bonaparte's Gull *Larus philadelphia*
  Very rare vagrant
Common Gull *Larus canus*
  Common winter visitor
Lesser Black-backed Gull *Larus fuscus*
  Small breeding population, increasing*
Herring Gull *Larus argentatus*
  Common breeding resident
Iceland Gull *Larus glaucoides*
  Rare winter vagrant
Glaucous gull *Larus hyperboreus*
  Rare winter vagrant
Great Black-backed Gull *Larus marinus*
  Common winter visitor
Kittiwake *Rissa tridactyla*
  Irregular passage migrant, mostly during NE gales
Sandwich Tern *Sterna sandvicensis*
  Common passage migrant
Roseate Tern *Sterna dougallii*
  Rare passage migrant
Common Tern *Sterna hirundo*
  Common passage migrant, summer breeding visitor to other parts of Swale
Arctic Tern *Sterna paradisaea*
  Scarce passage migrant
Little Tern *Sterna albifrons*
  Summer visitor, passsage migrant, attempts to breed most years
Black Tern *Chlidonias niger*
  Late summer passage migrant
Guillemot *Uria aalge*
  Occasional winter visitor
Razorbill *Alca torda*
  Rare winter visitor
Little Auk *Alle alle*
  Scarce winter visitor
Puffin *Fratercula arctica*
  Rare winter visitor
Feral Pigeon *Columba livia*
  Breeding resident
Stock Dove *Columba oenas*
  Breeding resident and migrant
Wood Pigeon *Columba palumbus*
  Common breeding resident
Collared Dove *Streptopelia decaocto*
  Common breeding resident
Turtle Dove *Streptopelia turtur*
  Common breeding visitor
Ring-necked Parakeet *Psittacula krameri*
  Occasional visitor
Cuckoo *Cuculus canorus*
  Common summer breeding visitor
Barn Owl *Tyto alba*
  Scarce, breeding resident
Little Owl *Athene noctua*
  Scarce breeding resident
Tawny Owl *Strix aluco*
  Scarce breeding resident
Long-eared Owl *Asio otus*
  Rare winter visitor

Short-eared Owl *Asio flammeus*
   Irregular winter visitor
Nightjar *Caprimulgus europaeus*
   Summer breeding visitor
Swift *Apus apus*
   Common summer breeding visitor
Alpine Swift *Apus melba*
   Rare visitor
Kingfisher *Alcedo atthis*
   Scarce winter visitor
Hoopoe *Upupa epops*
   Rare migrant
Green Woodpecker *Picus viridis*
   Resident breeding species
Great Spotted Woodpecker *Dendrocopos major*
   Resident breeding species
Lesser Spotted Woodpecker *Dendrocopus minor*
   Scarce breeding resident
Woodlark *Lullula arborea*
   Scarce migrant
Skylark *Alauda arvensis*
   Common breeding resident, passage and winter visitor
Shorelark *Eremophila alpestris*
   Rare winter visitor to coast
Sand Martin *Riparia riparia*
   Passage migrant and summer visitor
Swallow *Hirundo rustica*
   Common summer breeding visitor
House Martin *Delichon urbica*
   Common summer breeding visitor
Richard's Pipit *Anthus novaeseelandiae*
   Rare migrant
Tree Pipit *Anthus trivialis*
   Passage migrant, scarce breeding visitor
Meadow Pipit *Anthus pratensis*
   Common breeding resident, passage and winter visitor
Blyth's Pipit *Anthus godlewskii*
   Very rare autumn migrant
Rock Pipit *Anthus spinoletta*
   Regular winter visitor
Yellow Wagtail *Motacilla flava flavissima*
   Summer breeding visitor
Grey Wagtail *Motacilla cinerea*
   Occasionally seen in winter near freshwater
Pied Wagtail *Motacilla alba yarrelii*
   Breeding resident, passage migrant
Waxwing *Bombycilla garrulus*
   Rare winter visitor
Wren *Troglodytes troglodytes*
   Breeding resident

Dunnock *Prunella modularis*
   Common breeding resident
Robin *Erithacus rubecula*
   Breeding resident
Nightingale *Luscinia megarhynchos*
   Locally common summer breeding visitor
Black Redstart *Phoenicurus ochruros*
   Scarce winter visitor, passage migrant
Redstart *Phoenicurus phoenicurus*
   Uncommon passage migrant
Whinchat *Saxicola rubetra*
   Passage migrant
Stonechat *Saxicola torquata*
   Regular winter visitor
Wheatear *Oenanthe oenanthe*
   Common spring and autumn passage migrant
Ring Ouzel *Turdus torquatus*
   Annual but scare passage migrant
Blackbird *Turdus merula*
   Common resident
Fieldfare *Turdus pilaris*
   Common winter visitor
Song Thrush *Turdus philomelos*
   Resident and winter visitor
Redwing *Turdus iliacus*
   Common winter visitor
Mistle Thrush *Turdus viscivorus*
   Breeding resident
Cetti's Warbler *Cettia cetti*
   Rare vagrant
Grasshopper Warbler *Locustella naevia*
   Scarce summer visitor, breeds occasionally
Savi's Warbler *Locustella luscinioides*
   Rare summer visitor
Aquatic Warbler *Acrocephalus paludicola*
   Rare autumn migrant
Sedge Warbler *Acrocephalus schoenobaenus*
   Common breeding visitor
Marsh Warbler *Acrocephalus palustris*
   Rare migrant
Reed Warbler *Acrocephalus scirpaceus*
   Common breeding visitor
Icterine Warbler *Hippolias icterina*
   Rare vagrant
Desert Warbler *Sylvia nana*
   Rare vagrant
Barred Warbler *Sylvia nisoria*
   Rare migrant
Lesser Whitethroat *Sylvia curruca*
   Common breeding visitor
Common Whitethroat *Sylvia communis*
   Common breeding visitor

Garden Warbler *Sylvia borin*
  Regular breeding visitor
Blackcap *Sylvia atricapilla*
  Common breeding visitor, some overwinter
Yellow-browed Warbler *Phylloscopus inornatus*
  Rare autumn migrant
Wood Warbler *Phylloscopus sibilatrix*
  Rare passage migrant
Chiffchaff *Phylloscopus collybita*
  Common breeding visitor, some overwinter
Willow Warbler *Phylloscopus trochilus*
  Common breeding visitor
Goldcrest *Regulus regulus*
  Scarce breeding resident, passage migrant
Firecrest *Regulus igicapillus*
  Rare vagrant
Spotted Flycatcher *Muscicapa striata*
  Scarce breeding visitor, passage migrant
Pied Flycatcher *Ficedula hypoleuca*
  Autumn passage migrant
Bearded Reedling *Panurus biarmicus*
  Resident, small breeding population
Long-tailed Tit *Aegithalos caudatus*
  Common breeding resident
Coal Tit *Parus ater*
  Scarce breeding resident
Blue Tit *Parus caeruleus*
  Common breeding resident
Great Tit *Parus major*
  Common breeding resident
Nuthatch *Sitta europaea*
  Scarce breeding resident
Treecreeper *Certhia familiaris*
  Scarce breeding resident
Golden Oriole *Oriolus oriolus*
  Rare migrant
Great Grey Shrike *Lanius excubitor*
  Rare visitor
Jay *Garrulus glandarius*
  Common breeding resident
Magpie *Pica pica*
  Common breeding resident
Jackdaw *Corvus monedular*
  Common breeding resident
Rook *Corvus frugilegus*
  Common, scarce breeding resident
Carrion Crow *Corvus corone*
  Common breeding resident
Starling *Sturnus vulgaris*
  Common breeding resident and winter visitor
House Sparrow *Passer domesticus*
  Common breeding resident

Tree Sparrow *Passer montanus*
  Scarce resident? Winter visitor
Chaffinch *Fringilla coelebs*
  Common breeding resident, passage migrant, winter visitor
Brambling *Fringilla montifringilla*
  Scarce winter visitor, passage migrant
Greenfinch *Carduelis chloris*
  Common breeding resident, passage migrant, winter visitor
Goldfinch *Carduelis carduelis*
  Breeding resident, summer visitor, passage migrant
Siskin *Carduelis spinus*
  Passage migrant and winter visitor
Linnet *Carduelis cannabina*
  Breeding resident, passage migrant, summer and winter visitor
Twite *Carduelis flavirostris*
  Rare winter visitor
Redpoll *Carduelis flammea*
  Resident? Passage migrant, winter visitor,
Common Crossbill *Loxia curvirostra*
  Scarce winter visitor
Scarlet Rosefinch *Carpodacus erythrinus*
  Very rare migrant
Bullfinch *Pyrrhula pyrrhula*
  Breeding resident
Hawfinch *Coccothraustes coccothraustes*
  Rare, records only from Clowes Wood complex, resident?
Lapland Bunting *Calcarius lapponicus*
  Rare winter visitor
Snow Bunting *Plectrophenax nivalis*
  Regular winter visitor
Yellowhammer *Emberiza citrinella*
  Scarce breeding resident, declining
Reed Bunting *Emberiza schoeniclus*
  Breeding resident, passage migrant and winter visitor
Corn Bunting *Miliaria calandra*
  Scarce breeding resident and winter visitor

Whitstable Natural History Society Study Area

Species Index

A

Acrocephalus schoenobaenus 51
Acrocephalus scirpaceus 51
Alauda arvensis 45
Alcedo atthis 43
Anas crecca 14
Anas penelope 13
Anas platyrhynchos 17
Arctic Skua 5
Ardea cinerea 10
Arenaria interpres 34
Asio flammeus 42

B

Bar-tailed Godwit 5, 29
Bearded Reedling 52
Black-headed Gull 35, 36
Black-tailed Godwit 5, 28, 29
Black-throated Diver 7
Branta bernicla bernicla 11
Branta bernicla hrota 11
Brent Goose 5, 6, 11

C

Calidris canutus 26
Carduelis cannabina 45
Charadrius hiaticula 22
Circus cyaneus 19
Common Gull 36, 37
Common Tern 6, 41
Cormorant 9
Corn Bunting 55
Curlew 6, 30, 31

D

Dunlin 5, 22, 26, 27, 34

E

Eider 15
Emberiza schoeniclus 55

F

Falco tinnunculus 20

G

Gallinula chloropus 17
Gannet 5
Gavia stellata 7
Golden Plover 23, 24
Goldfinch 4, 45
Great Crested Grebe 8
Great Northern Diver 7
Great Skua 5
Great Black-backed Gull 38
Greenfinch 4, 45
Greenshank 5, 33
Grey Heron 10
Grey Plover 23, 24

H

Haematopus ostralegus 21
Hen Harrier 19
Herring Gull 36, 37

K

Kestrel 20
Kingfisher 43
Knot 5, 26

L

Lapwing 25
Larus argentatus 37
Larus canus 36
Larus marinus 38

Larus ridibundus 35
Leach's Petrel 5
Lesser Black-backed Gull 38
Limosa lapponica 29
Limosa limosa 28
Linnet 45
Little Tern 41
Long-tailed Skua 5

M

Mallard 17
Marsh Harrier 19
Meadow Pipit 5, 45, 46
Mergus serrator 18
Miliaria calandra 55
Moorhen 17
Motacilla flava flavissima 47

N

Numenius arquata 31
Numenius phaeopus 30

O

Oenanthe oenanthe 49
Oystercatcher 6, 21

P

Panurus biarmicus 52
Phalacrocorax carbo 9
Pied Flycatcher 5
Plectrophenax nivalis 53
Pluvialis apricaria 23
Pluvialis squatarola 24
Podiceps cristatus 8
Pomarine Skua 5

R

Red-breasted Merganser 18
Red-throated Diver 7
Redshank 5, 6, 32, 33
Reed Bunting 55

Reed Warbler 51
Ringed Plover 5, 6, 22, 34

S

Sabine's Gull 5
Sandwich Tern 6, 39
Saxicola torquata 48
Sedge Warbler 51
Shelduck 6, 13
Short-eared Owl 42
Skylark 5, 45, 46
Snow Bunting 53
Somateria mollissima 15
Sparrowhawk 20
Starling 4
Sterna albifrons 41
Sterna hirundo 41
Sterna sandvicensis 39
Stonechat 48
Swallow 6

T

Tadorna tadorna 13
Teal 14
Tringa nebularia 33
Tringa totanus 32
Turnstone 5, 22, 34

V

Vanellus vanellus 25

W

Wheatear 6, 49
Whimbrel 6, 30
Wigeon 13, 14

Y

Yellow Wagtail 6, 47

63

The Whitstable Natural History Society was formed in June 1973. From the start its aims have been to encourage an interest in the natural environment and show the benefits and importance of that environment. The Society meets at the Castle, Whitstable, from September to May and Field trips are organised throughout the year.

New members are always welcome. Telephone 01227 275936

The Whitstable Improvement Trust is an Independent Charitable Trust and Limited Company devoted to the careful preservation and the regeneration of Whitstable. It seeks to retain and care for the unique nature of the locality and its buildings, whilst creating an awareness of the town's historical past and the characters who have contributed to it. Initially funded by the city and county councils, the Trust now exists through its own resources and initiatives, with the support of its members.

Whitstable Improvement Trust
Harbour Street
Whitstable
Kent CT5 1AJ